Bibliographic information published by the German National Library:

The German National Library lists this publication in the National Bibliography; detailed bibliographic data are available on the Internet at http://dnb.dnb.de .

Imprint:

Copyright © 2012 GRIN Verlag, Open Publishing GmbH
Print and binding: Books on Demand GmbH, Norderstedt Germany
ISBN: 9783668386242

This book at GRIN:

http://www.grin.com/en/e-book/352028/tannin-binding-agents-comparison-of-the-effects-of-various-tannin-binding

Yisehak Kechero, Afework Bedewi, Assefa Getnet

Tannin binding agents. Comparison of the effects of various tannin binding agents on nutritive value of leaves of tropical tannin rich fodder trees

GRIN Publishing

Comparison of the effects of various tannin binding agents on nutritive value of leaves of tropical tannin rich fodder trees

Abstract

The study report here was undertaken to evaluate the nutritive value of leaves from tannin rich tropical multipurpose tree species with or without various tannin binding agents (TBAs). The studied species included *Albizia gummifera, Carissa edulis, Draceana steudneri, Ficus sycomorus, Grewia ferruginea, Millettia ferruginea, Prunus africana, Rhus glutinosa, Syzygium guineense and Ekebergia capensis.* Six independent samples per species were collected separately. The species were subjected to proximate, detergent and polyphenolic analysis. Dry matter and organic matter digestibility (DMD & OMD) of the plant leaves were determined following two stage in vitro enzymatic analysis. Dried plant species were incubated in two runs with three replications per run with rumen liquid collected from three Boran $\times$ Holstein-Friesian crossbred steers and gas production measured at 2, 4, 6, 8, 12, 24, 32, 48, 58, 72 and 96 h of incubation. The TBAs included were polyethylene glycol (PEG$_{4000}$ and PEG$_{6000}$), polyvinylpyrrilodone (PVP) and polyvinylpolypyrrilodone (PVPP). The total phenolics content (g/kg DM) ranged from 69 in G. *ferruginea* to 288 in C. *edulis*, which also had the higher total tannins (204 g/kg DM; P<0.001). The potential extent of gas production (b) was highest for *F. sycomorus* (249.83 ml), *E. capensis* (239.1 ml), and M. ferruginea (224.9 ml/g DM) as compared to other species (b$\leq$ 212; P<0.001). *F. sycomorus* and *R. glutinosa* had the highest gas production rate (c; 10.2 and 10.9 ml/h, respectively (P<0.001) compared to the other species. Treatments receiving PEG$_{6000}$ and PEG$_{4000}$ were found to be superior over other treatments in terms of extent and rate of gas production (P<0.001). Total tannin (TT) and condensed tannin (CT) contents were negatively correlated to *in vitro* gas production $\geq$ 12h (P<0.05). Similarly, TT and CT contents were negatively correlated to DMD, and OMD (r = -0.53 to -0.72; P <0.0.01). Tannin binding agents have different effects on nutritive value among species; yet, the effect size of TBAs is not related to their respective tannin contents. Generally, addition of PEG$_{6000}$, PEG$_{4000}$, PVP and PVPP could help to overcome adverse effects of tannins on nutrient availability as indicated by *in vitro* gas production and estimated parameters. Based on their nutrient content and fermentation kinetics after addition of TBAs, the browse species have a good potential as a sustainable livestock feed resources.

Key words: Browse species, in vitro gas production, tannin, tannin binding agent, short chain fatty acid, metabolisable energy

Content

Introduction

The major constraint to increase livestock productivity in Sub-Saharan African countries and many other developing nations is the scarcity and fluctuating quantity and quality of the year-round supply of feed resources (Besharati and Taghizadeh, 2009; Arigbede et al., 2011). The native pastures and crop residues are the major feed resources available in these areas for livestock (Salem et al., 2006). However, tropical graminaceous fodder and crop by-products have a low nutritive value due to their low protein and fermentable energy content (El Hassan et al., 2000). Addition of foliages of multipurpose tree species (MPTS) are of importance in animal production because they can provide a significant amount of protein, especially in the dry season and can improve the utilization of low quality roughages mainly through the supply of protein to rumen microbes (Makkar, 2003; Ozturk et al., 2006).

Although MPTS have important nutritional merits (Kaitho et al., 1998; Norton, 2000), there are also reports which indicate that secondary plant metabolites are found in many fodder trees and shrubs such as condensed tannins (CTs) (Makkar and Becker, 1998; Frutos et al., 2004) that can affect feed intake, digestibility, growth, onset of puberty and reproductive functions via interference in the nutrient availability.

In vitro gas production has been used to assess the nutritive value of browse species. It basically results from fermentation of carbohydrates, proteins and other organic compounds (Blümmel and Ørskov, 1993). Hence, the extent and rate of gas production reflects the efficiency and/or extent of degradability of feedstuffs.

As the critical demand for livestock feed rises particularly in smallholder subsistence farming systems in many developing countries, the new tannin rich MPTS could make up an increasingly important part of livestock diets in addition to sources of protein for strategic supplementation. It is therefore critical that techniques be developed to measure and manage the tannins that they contain. Addition of tannin deactivating agents (TBAs) (Silanikove et al., 2001; Makkar, 2003; Hagerman, 2011) such as polyethylene glycol (PEG$_{4000}$ & EG$_{6000}$), polyvinylpyrrolidone (PVP) and polyvinylpolypyrrolidone (PVPP) would be efficient in decreasing and/or neutralizing their negative effects and improving nutrient utilization. All tannin binding compounds are commercially available in a range of various molecular weights and using them in small quantities would improve their purchasing prices. It has been suggested that occurrence of tannins and their effect on extent and rate of *in vitro* fermentation from tannin rich browses can be assessed using gas production techniques coupled with the use of various tannin-binding agents.

This study is the first to compare the effect of four different tannin-binding agents on the nutritive value of leaves of ten tree species identified as potential feed resources. Information on rate and extent of *in vitro* fermentation of this type of feed resources is lacking. Evaluating *in vitro* fermentation kinetics in relation to nutrient and polyphenolics composition can help in estimating the feeding value of previously untested MPTS. The objectives of the present study were to determine the chemical composition of a range of tannin rich multipurpose tree spp., to evaluate and quantify their *in vitro* fermentation characteristics, to compare the efficacy of tannin-deactivating agents on *in vitro* fermentation characteristics, and to examine the relationship between the chemical composition and *in vitro* fermentation parameters of the tested feed resources.

Materials and methods

Study site

The chemical composition and polyphenolics analysis were carried out in the Animal Nutrition Laboratory of Jimma University, College of Agriculture and Veterinary Medicine, Jimma, south western Ethiopia. Jimma is situated at 7°40′N and 36°50′E latitude and longitude, respectively at an altitude of 1780 meters above sea level (m.a.s.l). The *in vitro* gas production measurement was carried out in Holeta Agricultural Research Center (HARC) animal nutrition laboratory located in the West Shewa administrative zone of the Oromia National Regional State, Central Ethiopia, (latitude and longitude: 9° 3′ N, and 38° 30′ E and 35 km west of Addis Ababa, Ethiopia). The research centre is located at an average altitude of 2391 masl.

Leaves of the plant species

Leaves of 10 browse plant species, rich in condensed tannins (CT), widely known, plentiful and highly preferred by local farmers for their multiple uses (Yisehak et al., 2010a), were sampled from the Wakitola farmers association (7°71' N – 37°04' E, 1718 masl) of the Omo Nada district, Jimma administrative zone, South western Ethiopia. The leaves of the plant species (mixture of immature, young and old, this is because practically ruminants consume mixture of all leaf types) were collected from those study sites described in chapter 5. According to livestock farmers, the age of fodder trees was about 5 years. The collected leaves were put in plastic bags and directly placed on ice in an insulated container and transported under dark conditions to the laboratory within 30 minutes after collection.

The plant species studied were *Albizia gummifera*, *Carissa edulis*, *Dracaena steudneri*, *Ficus sycomorus*, *Grewia ferruginea*, *Millettia ferruginea*, *Prunus africana*, *Rhus glutinosa*, *Syzygium guineense* and *Ekebergia capensis*. Leaves of six individual plants per species were sampled from the same locality having similar climatic and soil characteristics and analyzed individually to allow quantification of the intra-species variation and to perform statistical analyses in a factorial procedure. Plant samples were replicated in order to obtain sufficient number of statistical units (Robinson et al., 2006). Based on FAO/IAEA (2000) recommendations, immediately after arrival at the laboratory, the samples were dried at approximately 55°C to constant weight using a forced air oven. After drying, samples were ground and stored in air tight containers in dark cabinets at room temperature (avg. 20°C) until laboratory analyses.

Chemical analyses and nutritive value

Dry matter (DM) was determined by drying the leaf samples at 105°C overnight, crude ash (CA) by incinerating the samples at 550 °C for 8 h, ether extract (EE) using anhydrous diethyl ether, and nitrogen (N) by the Kjeldahl method (AOAC, 2005 976.05, using an automated steam distillation/titration unit (FOSS – Kjeltec 2300 Analyzer) with 1% boric acid as the receiving solution and 0.1 M hydrochloric acid as the titrant, with crude protein (CP) concentration calculated as N×6.25. Neutral detergent fiber (aNDF-NDF) and acid detergent fiber (ADF-ADF) of leave samples were determined by the method of Van Soest et al. (1991) in an ANKOM 200 Fiber Analyzer (ANKOM Technologies, Macedon, NY). Lignin was determined by the sulfuric acid method (lignin (sa)) (Robertson and Van Soest 1981) in the ANKOM 200 fiber analyzer. Neutral detergent fiber insoluble/bound nitrogen (aNDF-NDFIN) is determined as the N in aNDF-NDF residue (non-ashed) following the Kjeldahl method and the CP concentration was calculated N×6.25. All chemical constituents are reported on a DM basis.

The total carbohydrate (CHO, g/kg DM) and non-structural carbohydrate (NSC, g/kg DM) concentrations were estimated according to Ranjhan (1997; eq. 1 & 2):

Total CHO = 1000 – (CP + EE + Ash + lignin; all components in g/kg DM)............[1]

Non-structural carbohydrate (NSC, g/kg DM) concentrations were calculated as described by McDonald et al. (2002):

$$\text{NSC} = 1000 - (\text{CP} + \text{ash} + \text{EE} + (\text{aNDF-NDF minus NDFCP})) \dots\dots\dots\dots\dots\dots\dots [2]$$

Where, NSC accounts for the NDF bound crude protein (NDFCP) and avoids double counting of some CP.

The two stage *in vitro* technique developed by Tilley and Terry (1963) was used to determine dry matter and organic matter digestibility (DMD & OMD) as modified by Van Soest and Robertson (1985), in which the second stage (pepsin digestion) is substituted with neutral detergent extraction. This treatment removes all the indigestible microbial matter and leaves a residue of undigested plant cell wall and the values are true digestibility. The modified procedure is as precise and requires half the time to complete the whole procedure (two days shorter) compared to Tilley and Terry system. About 0.5 g of ground sample (1 mm sieve) was weighed into a test tube. A 10 ml rumen liquor and 50 ml buffer solution was added to the sample in the tube. The mixture was incubated at 39°C for 48 h, ensuring that it was carefully shaken from time to time. Finally, the tubes were centrifuged and the supernatant decanted. The residue was again incubated with 60 ml pepsin-hydrochloric acid solution (to digest protein) for another 48h at 39°C. This was followed by centrifugation, filtering, drying the residues and ashing. Two blanks (rumen liquor mixed with buffer only) and two standards with known digestibility were included to correct for the indigestible DM from the rumen liquor and to check whether the system was working perfectly. Metabolisable energy (ME, MJ/kg) value was estimated from the two stage *in vitro* organic matter digestibility: ME = 0.16 × OMD% (McDonald et al 2002).

Samples were also analyzed for total phenolics (TP), total tannins (TT) and condensed tannins (CT)(FAO/IAEA, 2000). TP (tannin phenolics and non-tannin phenolics) and TT were determined with the Folin-Ciocalteau reagent method. Insoluble matrix, polyvinyl polypyrrolidone (PVPP; as this polymer binds strongly to tannin-phenolics) was used for separating tannin phenolics from non-tannin phenolics. A calibration curve was prepared using a tannic acid standard. Total tannins (tannic acid equivalent) were estimated from the difference between total phenol and non-tannin phenol obtained after precipitating with polyvinyl polypyrrolidone (*P*-107302, MERCK chemicals Co. Ltd.). Both TP and TT values were calculated as tannic acid equivalent and expressed as tannic acid eq-g/kg DM. The absorbance for TP and TT was read at 725 nm using a spectrophotometer (Makkar et al., 1993).

The CT was measured by the butanol-HCl-Fe method and the results were expressed as leucocyanidin equivalent and the concentration of CT was calculated (% DM) from the formula [Abs.550nm $\times$ 78.26 $\times$ dilution factor/% DM] (Porter et al., 1986).

The NDF-bound CTs or PAs were determined in the dried neutral detergent fiber fraction of the browse plants after extraction of NDF (Van Soest et al., 1991). The non-ashed NDF residue samples of 200 mg were weighed into each of triplicate screw cap test tubes and 5 ml of butanol-HCl reagent added to each tube. In soluble NDF-bound proanthocyanidin (NDF-bPA) is expressed as Abs. 550nm, absorbance units (1 cm pathlength) per gram of NDF as described by Reed et al. (1982).

In vitro gas production measurement

Rumen fluid was obtained from three rumen-fistulated Boran (*Bos indicus*) $\times$ Holstein Friesian (*Bos taurus*) cross-bred steers, age of 9.4 years, fed twice daily with a diet containing pasture hay (60%) and concentrate mixture (40%) who had free access to water and mineral/vitamin licks. The ration was assumed to have meeting all their nutrient and energy requirements. Rumen liquid samples were collected before the morning meal in pre-warmed thermos flasks and transported immediately to the adjoining laboratory where it was strained through four layers of cheesecloth and kept at 39°C under a CO_2 atmosphere. Filtered rumen liquid was pooled together in order to achieve homogenous inocula.

To prepare the buffer solution and rumen inocula the procedure of Theodorou et al. (1994) was followed with rumen liquid and buffer mixed in the ratio 1:2 (v/v). In short, about 300 mg DM of ground substrates was weighed in triplicate into serum bottles kept at approximately 39°C and flushed with CO_2 before use.

At the start of incubation, bottles were inoculated with 60 ml of buffered rumen liquid, closed airtight with rubber stoppers, crimp sealed and placed into an incubator set at 39°C, being shaken at every 2 h. The volume of gas produced in each bottle was recorded at successive incubation times at 2, 6, 8, 12, 24, 32, 48, 72 and 96 h after the start of the incubation using the pressure transducer technique (Theodorou et al. 1994). Incubations were performed in two runs with three replications per run. Three blanks and a standard hay sample of known gas production were included in each run. The values of pH were measured from the fermenting substrates at the end of incubation. All substrates were also incubated in the presence of different tannin-binding agents (TBAs; PEG_{4000}, PEG_{6000}, PVP & PVPP). All TBAs were products of MERCK Chemicals Co. Ltd, 2010-2012. Dosage selection of TBAs was based on the recommendations of Silanikove et al. (2001), Makkar et al. (1995) and Makkar (2003). Accordingly, the levels can range from 0% to equivalent DM weight substrates. Each TBA was dissolved in the buffer solution (NH_4HCO_3 and $NaHCO_3$) to give a concentration of 0, 0.5 and 1.0g TBA per gram DM of sample.

Total gas productions were corrected within batch for the corresponding blank across batches following the hay standards with known gas production. The kinetics of gas production data were obtained after fitting the modified exponential equation model to the data points, where the the assumption that gas production from the soluble fraction (ml)/the intercept (a) was not included in the model with the understanding that no gas was produced from unfermented feed (Ørskov and McDonald, 1979; Eq. 3).

Thus, the original equation $GP = a + b\ (1\text{-}e^{-ct})$ was modified to $GP = b\ (1\text{-}e^{-ct})$.................[3]

Where: GP represents the volume of gas produced(ml) with time (t); b, the potential extent of gas production (ml); c, the gas production rate constant for b (ml/h); e, the base for natural logarithms and t, incubation time.

Statistical analyses

For the gas production parameters and their derivates, a variance analysis model with three fixed factors was used: plant species (P; 10 levels), tannin binding agents (T; 4 levels) and dosage of tannin binding agents (D; 3 levels). The linear model used was: $Y_{ijk} = \mu + P_i + T_j + D_k + (P{\times}T)_{ij} + (P{\times}D)_{ik} + (T{\times}D)_{jk} + (P{\times}T{\times}D)_{ijk} + \varepsilon_{ijk}$; where, Y_{ijk} is the observation of the ith plant species, j^{th} tannin binding agent and k^{th} doses of the tannin binding agent; μ, the population mean, P_i, i^{th} plant species effect (i = 1 to 10); T_j, the jth effect of tannin binding agents (j = 1 to 4); D_k, the kth effect of dose of the tannin binding agents; $(P{\times}T)_{ij}$, is the ij^{th} interaction effect between plant species and tannin binding agent; $(P{\times}D)_{ik}$, the ik^{th} interaction effect between plant species and doses of tannin binding agents; $(T{\times}D)_{jk}$, jk^{th} interaction effect between tannin binding agents and doses of the agents; $(P{\times}T{\times}D)_{ijk}$, ijk_{th} interaction effect between plant species, tannin binding agents and doses of tannin binding agents and ε_{ijk}, the residual error.

For all other parameters, the model $Y_{ij} = \mu + P_i + \varepsilon_{ij}$ was used for plant species evaluated without tannin binding agents.

The non-linear fermentation parameters such b, c, and t were estimated for each plant species and tannin binding agent combination using nonlinear (NLIN) procedures of SAS. Multiple run values were averaged before being analyzed and not included in the statistical model. All analyses were performed with SAS (2010 version 9.3). Differences between means were tested using Tukey test with significance declared at $P<0.05$. Standard errors of means were calculated from the residual mean square in the analysis of variance. A Pearson correlation analysis was used to establish the relationship between chemical composition and *in vitro* gas production parameters.

Results

Chemical composition, polyphenolic components and in vitro digestibility of the tannin rich plant leaves

The chemical composition, *in vitro* digestibility and estimated nutritive value parameters of the leaves of the tannin rich tree species are presented in Table 1. The concentration of polyphenolics varied widely between plant species ($P<0.001$). The ash content (g/kg DM) ranged from 28 and 112. Highest mean values for CP (g/kg DM) were recorded for *A. gummifera* (299) ($P<0.001$). Interspecies variation was also observed for the concentration of NDF-CP ($P<0.001$). The highest OMD values were recorded for *D. steudneri* (667 g/kg DM) and *P. africana* (634 g/kg DM) ($P<0.001$). The ME contents were recorded for these two plants (*D. steudneri* and *P. africana*) were also found to be the highest (10.7-10.20 MJ/kg DM), which was significantly different from all other species ($P<0.001$). The lowest mean CT concentration (g/kg DM) was found for *G. ferruginea* (53), whereas the highest CT value was observed in three species (*C. edulis*, 175; *R. glutinosa*, 190; *S. guineense*, 166; leucocyanidin tannin equivalent). *C. edulis*, and *G. ferruginea* had the largest and smallest TP concentration, 288 and 69 g /kg DM, respectively ($P<0.001$). The NDF-bPA content varied between plant species, the highest value recorded was 24 absorbance units (AU)/g (*A. gummifera*) and the two lowest 7 and 8 AU/g for *G. ferruginea* and *D. steudneri*, respectively.

Table 1. Chemical composition (g/kg DM), *in vitro* digestibility (g/kg DM), metabolisable energy (MJ/kg) and polyphenolic concentration (g/kg DM) of leaves tannin-rich trees species (N = 6)

Nutriments	Albizia gummifera	Carissa edulis	Draceana steudneri	Ficus sycomorus	Grewia ferruginea	Millettia ferruginea	Prunus Africana	Rhus glutinosa	Syzygium guineense	Ekebergia capensis	SEM	P-value
DM	932[bc]	905[d]	921[dc]	939[ab]	942[ab]	951[a]	908[d]	932[bc]	913[d]	941[ab]	4	<0.001
Ash	81[g]	92[d]	103[b]	112[a]	97[c]	88[e]	97[c]	83[gf]	84[f]	28[h]	4	<0.001
CP	299[a]	138[f]	218[d]	173[e]	229[c]	242[b]	139[f]	145[e]	131[g]	144[f]	10	<0.001
NDF-CP	16[a]	10[g]	13[d]	12[e]	14[c]	15[b]	11[f]	11[f]	10g	11[f]	0.1	<0.001
EE	45[b]	40[dc]	24[g]	21[h]	29[f]	32[e]	48[a]	41[c]	39[d]	15[i]	2	<0.001
aNDF-NDF	394[bc]	370[bc]	380[bc]	416[bc]	350[bc]	415[bc]	452[ab]	408[bc]	554[a]	323[c]	19	<0.001
ADF-ADF	283[c]	287[c]	328[bc]	348[bc]	317[bc]	380[b]	331[bc]	330[bc]	465[a]	192[d]	16	<0.001
NSC	194[d]	370[ce]	288[bcd]	290[bcd]	309[bcd]	238[cd]	275[bcd]	334[bc]	202[d]	501[a]	6	<0.001
Lign	121[ab]	84[c]	107[bc]	73[c]	110[bc]	148[a]	101[bc]	80[c]	127[ab]	78[c]	7	<0.001
DMD	348[e]	355[e]	649[a]	424[c]	423[d]	491[c]	621[b]	318[f]	360[e]	431[d]	9	<0.001
OMD	399[d]	394[d]	667[a]	470[c]	541[b]	508[bc]	634[a]	363[d]	388[d]	513[bc]	20	<0.001
CHO	453[c]	646[b]	548[c]	621[bc]	535[c]	491[d]	615[b]	652[b]	619[bc]	736[a]	11	<0.001
ME	6.4[d]	6.25[d]	10.7[a]	7.48[c]	8.7[b]	8.13[b]	10.2[a]	5.78[e]	6.1[d]	8.15[b]	2	<0.001
CT	79[de]	175[a]	65[f]	113[c]	53[g]	73[e]	76[de]	171[ab]	166[b]	82[d]	33	<0.001
TT	86[f]	204[a]	77[g]	124[d]	59[h]	76[g]	109[e]	190[b]	175[c]	120[d]	9	<0.001
TP	118[g]	288[a]	101[i]	154[f]	69[j]	105[h]	157[e]	240[b]	236[c]	171[d]	13	<0.001
NDF-bPA	24[a]	12[e]	8[g]	10[f]	7[h]	22[b]	10[f]	13[d]	15[c]	11[e]	0.4	<0.001

[a,b,c,d,e,f,g,h,i]*Means with different superscripts within the same row are significantly different DM, dry matter ; CP, crude protein ; NDF-CP , neutral detergent bound crude protein; EE, ether extract; aNDF-NDF, neutral detergent fibre assayed with a heat stable amylase and expressed inclusive of residual ash; ADF-ADF, acid detergent fibre expressed inclusive residual ash; NSC, nonstructural carbohydrate; Lign(sa), lignin determined by solubilization of cellulose with sulphuric acid; DMD, OMD, in vitro organic matter digestibility; CHO, total carbohydrates; ME, metabolisable energy; CT, total extractable/soluble condensed tannin, TT, Total extractable tannin; TP, total extractable phenols; NDF-bPA, NDF-bound Proanthocyanidin (AU/g); SEM, standard errors of means; ***P<0.001*

11

In vitro gas production potentials of the plant leaves

Interspecies variation for *in vitro* gas production characteristics is reported in Table 2. In the initial incubation time (2 h) the highest and lowest gas volume was recorded for *E. capensis* (21 ml) and *A. gummifera* (1 ml) ($P<0.001$). Although interspecies variation was significant ($P<0.001$) for gas production across all the incubation times, the gas volumes over 100 ml were recorded starting from 12 hours of incubation ($P<0.001$). However, *S. guineense* had a gas production over 100 ml only after 48h of incubation ($P<0.001$). At 96 hours of incubation, the highest gas volumes were recorded for *F. sycomorus* (260 ml), *M. ferrugenia* (237 ml), *R. glutinosa* (233 ml), *E. capensis* (252 ml) and *D. steudneri* (234 ml) whereas the lowest gas volumes were recorded for *S. guineense* (157 ml) and *C. edulis* (164 ml)($P<0.001$).

Table 2. Effect of interspecies variation on *in vitro* gas production at different incubation times

Plant species	Incubation times (h, mean gas (ml/g DM)								
	2	6	8	12	24	32	48	72	96
Milletia ferrugenia	13[b]	32[bc]	53[bc]	77[c]	114[bc]	145[bc]	183[bcd]	201[bc]	237[ab]
Ficus sycomorus	16[b]	36[ab]	52[bcd]	75[c]	115[bc]	151[a]	189[abc]	227[ab]	260[a]
Rhus glutinosa	16[b]	34[b]	52[bcd]	101[ab]	122[b]	141[bc]	167[cde]	196[c]	233[ab]
Grewia fruginea	15[b]	31[bc]	50[cde]	66[c]	97[dc]	122[dc]	156[de]	185[c]	209[bc]
Ekebergia capensis	21[a]	40[a]	56[b]	83[bc]	133[ab]	173[a]	211[a]	242[a]	251[a]
Carissa edulis	17[b]	28[c]	42[f]	62[c]	85[d]	109[de]	129[gf]	152[de]	164[d]
Syzygium guineense	16[b]	31[bc]	45[ef]	65[c]	79[d]	92[e]	110[g]	131[e]	157[d]
Albizia gummifera	1.0[c]	21[d]	49[cde]	80[c]	113[bc]	144[bc]	168[cde]	186[c]	219[c]
Prunus africanus	15[b]	31[bc]	46[def]	68[c]	97[dc]	121[dc]	152[ef]	175[cd]	200[c]
Dracaena steudneri	15[b]	39[a]	69[a]	106[a]	145[a]	173[a]	203[ab]	226[ab]	234[ab]
SEM	0.57	0.68	0.88	2.45	2.93	3.44	3.9	4.1	4.15
P-value	<0.001	<0.001	<0.001	<0.001	<0.001	<0.001	<0.001	<0.001	<0.001

[a,b,c,d,e,f,g]*Means with the same letter in the columns are not significantly different; h, hour; SEM, standard error of means; ***P<0.001*

Effects of tannin binding agents on in vitro gas production and kinetics of the leaves

In the leaves of all plant species, tannin deactivation through tannin-binding agents (TBAs) significantly increased the cumulative gas production as compared to all treatments not incubated with TBAs (*GP*; $P<0.001$, Table 3). The largest increase was measured for PEG treatment groups compared to the controls ($P<0.001$). Among the treatment levels (doses), 1g of both PEG$_{6000}$ and PEG$_{4000}$ treatments were found to be superior for their effect on *in vitro* gas production compared to the other tannin-binding agents ($P<0.001$).

Table 3. Effects of tannin-binding agents on cumulative in vitro gas production of tannin-rich tree species (ml/g DM)

Plant species	Control	PEG$_{4000}$ 0.5g	1.0g	PEG$_{6000}$ 0.5g	1.0g	PVP 0.5g	1.0g	PVPP 0.5g	1.0g	SEM	P-value P×T×D
Albizia gummifera	219[d]	255[c]	337[b]	247[c]	434[a]	253[c]	254[c]	262[c]	267[c]	5	***
Catha edulis	164[f]	314[b]	355[a]	264[c]	251[c]	209[ed]	222[d]	197[e]	194[e]	6	***
Dracaena steudneri	234[e]	351[b]	391[a]	293[a]	393[a]	274[d]	337[b]	272[d]	297[c]	3	***
Ekebergia capensis	251[e]	331[b]	368[a]	331[b]	358[ab]	286[cd]	302[c]	253[cd]	271[d]	5	**
Grewia ferruginea	209[f]	330[c]	378[b]	354[bc]	404[a]	303[d]	306[d]	259[e]	256[b]	5	***
Milletia ferruginea	237[e]	309[d]	430[ab]	396[bc]	460[a]	270[de]	284[d]	312[d]	372[c]	6	***
Prunus Africana	200[f]	329[b]	381[ab]	315[b]	385[a]	267[cde]	283[abc]	237[e]	243[de]	5	***
Rhus glutinosa	233[d]	409[a]	334[a]	335[b]	324[ab]	223[d]	275[c]	251[de]	264[c]	5	***
Syzygium guineense	157[e]	275[b]	337[a]	292[b]	344[a]	220[cd]	233[c]	205[d]	200[d]	2	***
Ficus sycomorus	260[g]	401[c]	416[b]	420[b]	545[a]	340[ed]	351[d]	324[f]	337[ef]	5	***

PEG$_{4000}$, Polyethylene glycol 4000; PEG$_{6000}$, Polyethylene glycol 6000; PVP, polyvinylpyrrolidone; PVPP, polyvinylpolypyrrolidone; P×T×D, interaction of plants (P), tannin binding agents (T) and dosage of tannin binding agents (D); [a,b,c,d,e,f,g] Means with the different superscripts in rows are significantly different (P<0.05); SEM, standard error of means

Fermentation characteristics of tannin rich tree leaves

The *in vitro* gas production characteristics of the leaves of tannin-rich tree species incubated without TBAs are presented in Table 4. Interspecies variation was significant for the potential extent of gas production *b*- ($P<0.001$). The largest amount of gas produced (249.8 & 239.1 ml/g DM) from *b* was recorded for *F. sychomorus* and *E. capensis*, respectively) while the smallest gas production was recorded for *C. edulis* (148.1 ml/g DM) and *S. guineense* (132.0 ml/g DM) ($P<0.001$). The gas production rate *c* ranged between 10.9 ml/h in *R. glutinosa* to 3.2 ml/h in *M. ferruginea* and their volume of gas produced at time t (*GP, gas production*) was highly variable ($P<0.001$). Likewise, incubation period t varied between 51.6 h for *P. africana*, at which time maximum rate of gas production is reached and 19.0 h for *A. gummifera* ($P<0.001$).

Table 4. Fermentation characteristics of leaves of the tannin-rich tree species without tannin binding agent (defined by the equation: $GP = b (1-e^{-c\,t})$)

Parameter	*Albizia* *gummifera*	*Carissa* *edulis*	*Dracaena* *steudneri*	*Ficus* *sycomorus*	*Grewia* *ferruginea*	*Millettia* *ferruginea*	*Prunus* *africana*	*Rhus* *glutinosa*	*Syzygium* *guineense*	*Ekebergia* *capensis*	SEM	*P-value*
b, ml	186[e]	148[f]	198[de]	250[a]	188[e]	225[b]	190[de]	212[cd]	132[f]	239[ab]	7.29	<0.001
c, ml/h	7[b]	7.4[b]	5.8[c]	10.2[a]	5.5[c]	3.2[d]	5.5[c]	10.9[a]	3.5[d]	5[c]	0.34	<0.001
t, h	19[g]	39.4[cd]	23.4[fg]	30.4[ef]	49.8[ab]	23.4[fg]	51.6[a]	27.1[ef]	32.7[de]	43.8[bc]	2.59	<0.001
pH	6.48	6.72	6.37	6.47	6.86	6.43	6.78	6.64	6.56	6.79	0.05	NS
R^2	0.872	0.903	0.898	0.9	0.894	0.922	0.833	0.791	0.907	0.944	0.11	<0.001

b, potential extent of gas production (ml); c, rate constant of gas production during incubation (ml/h); t, fermentation time in hours; pH, measure of acidity or alkalinity; R^2, regression coefficient; [a,b,c,d,e,f,g] *Means with different superscripts in the rows are significantly different (P<0.05); SEM, Standard error of mean; NS, non-significant; *P<0.05; **P<0.001*

Effect of tannin binding agents on in vitro fermentation characteristics of tannin rich tree leaves

Tannin rich tree leaves incubated with TBAs significantly improved the extent of *in vitro* gas production as compared to leaves incubated without TBAs ($P<0.001$). Among the TBAs, both types of PEG (PEG_{4000} & PEG_{6000}, Table 5) significantly improved the potential extent of gas production (*b*) from tree leaves ($P<0.001$). The interaction effect between leaves of plant species (P), TBAs (T) and dosage of TBAs (D) had a significant effect on the potential extent of gas production ($P<0.001$).

Incubation of tannin rich leaves with and without TBAs showed a significant variation in the rate of gas production *c* between species ($P<0.001$). The PEG treatment also gave the highest *c* value for most plant species compared to other TBAs ($P<0.001$).

Substantial variation was also observed on the time (*t*) required to reach the maximum gas production rate between controls and those receiving TBAs ($P<0.001$). For all species, the longest time was recorded for maximum gas production rate after receiving TABs compared to control ($P<0.001$). Similar to *b* and *c*, the interaction effect $P \times T \times D$ was found to be significant for *t* ($P<0.001$).

Table 5. Effect of plant species (P), tannin binding agents (T) and dosage of tannin binding agents (D) on the kinetics of *in vitro* gas production

		Treatment										*P*-value
			PEG$_{4000}$		PEG$_{6000}$		PVP		PVPP			
		Control	0.5g	1.0g	0.5g	1.0g	0.5g	1.0g	0.5g	1.0g	SEM	P×T×D
Albizia	b	186^e	225^{cd}	285^b	231^c	396^a	226^{cd}	210^d	211^d	211^d	4.55	<0.001
gummifera	c	7^a	7.2^b	7.8^{ab}	7.5^{ab}	8.4^a	7.2^b	7.2^b	7.2^{ab}	7.4^{ab}	0.17	<0.001
	t	19^d	25^c	34^b	26^c	39^a	21^c	22^{cd}	23^{cd}	24^c	1.52	<0.001
Carissa	b	148^e	250^c	300^b	239^b	321^a	181^d	181^d	178^d	180^d	2.64	<0.001
edulis	c	7.4^c	8.9^b	9.3^b	9^b	10.4^a	8.9^b	8.9^b	8.7^b	8.9^b	1.15	<0.001
	t	39^d	48^{cd}	53^b	50^c	67^a	50^c	53^b	50^c	53^b	3.07	<0.001
Dracaena	b	198^f	302^c	366^b	306^b	378^a	252^{de}	271^d	235^e	247^{de}	2.35	<0.001
steudneri	c	5.8^d	8.4^b	13^b	9.3^b	14.6^a	8.3^c	9.2^b	9^b	9.1^b	0.63	<0.001
	t	23.4^c	29.2^b	37.4^b	30^c	42.3^a	29.2^c	32.2^{bc}	29.3^c	30.5^c	3.4	<0.001
Ekebergia	b	239^f	300^c	356^b	303^c	379^a	247^e	257^d	245^e	253^d	4.01	<0.001
capensis	c	5^e	8.1^c	11.4^b	8^c	13.2^a	6.4^d	6.7^d	6.8^d	7.2^{cd}	1.12	<0.001
	t	44^e	53^c	60^b	54^c	65^a	46^d	54^c	51^{cd}	51^{cd}	1.89	<0.001
Grewia	b	188^e	269^b	311^b	268^b	324^a	223^d	245^c	238^{cd}	243^c	4.22	<0.001
ferruginea	c	5.5^e	9^c	10.5^b	9.1^c	13.1^a	7.3^d	7.9^d	7.7^d	8.2^{cd}	1.41	<0.001
	t	50^{cd}	58^b	71^a	56^a	72^a	53^c	56^{bc}	53^c	59^b	0.5	*
Milletia	b	225^f	285^d	379^b	320^c	400^a	229^f	265^{de}	250^e	330^c	4.02	<0.001
ferruginea	c	3.2^e	5.5^{bc}	11.8^a	5.8^b	12^a	3.8^{de}	4^d	3.9^{de}	5.2^{bc}	1.11	<0.001
	t	23.4^d	28.3^c	35.5^a	28^c	35.5^a	28^c	27^c	29^c	33.3^b	0.8	<0.001

Prunus	b	190[f]	280[c]	322[b]	281[c]	355[a]	256[d]	277[c]	250[e]	267[d]	3.52	<0.001
aricana	c	5.5[f]	8.1[d]	10.2[b]	8.3[bc]	11.7[a]	7.1[e]	7.5[e]	8.5[d]	9.7[c]	0.7	<0.001
	t	51.6[f]	75[d]	86.3[b]	79.1[c]	88.2[a]	70.2[e]	72.1[de]	74.2[d]	74.8[d]	2.01	<0.001
Rhus	b	212[e]	366[c]	415[b]	370[b]	455[a]	220[d]	231[d]	225[d]	237[d]	3.12	<0.001
glutinosa	c	10.9[e]	12.2[c]	14.1[b]	11.7[d]	15[a]	11.5[d]	11.5[d]	12.1[c]	11.5[d]	1.3	<0.001
	t	27.1[ab]	32[ab]	38[a]	33[a]	41[a]	30.1[a]	31.7[a]	30.5[a]	30.8[a]	0.93	0.03
Syzygium	b	132[g]	229[d]	268[b]	240[c]	285[a]	200[e]	211[e]	179[f]	195[e]	1.32	<0.001
guineense	c	3.5[f]	4.9[c]	5.11[b]	4.7[d]	6.1[a]	4.1[d]	4.3[d]	4[e]	4.4[d]	1	<0.001
	t	33[e]	40[c]	46[ab]	43[b]	51[a]	35[d]	37[d]	35[d]	39[c]	0.31	0.01
Ficus	b	250[d]	325[c]	371[b]	345[bc]	412[a]	270[b]	293[d]	300[cd]	320[c]	2.44	<0.001
sycomorus	c	10.2[c]	12[b]	15[ab]	13[b]	17[a]	11[b]	12[b]	12[b]	13[b]	1.52	<0.001
	t	30.4[c]	33[b]	42[a]	33[b]	44[a]	31[b]	32[b]	30[bc]	32[b]	1.05	<0.001

*b, the potential extent of gas production (ml); c, rate constant of gas production during incubation (ml·h⁻¹); t, fermentation time to reach maximum rate of gas production (h); [a,b,c,d,e,g,h]Means with superscripts in the rows are significantly different (P<0.001); SEM, Standard error of mean; P, plant species;T, TBA; D, dose; *P<0.05; ***P<0.001*

Correlation between some chemical composition and in vitro gas production

The correlation coefficient between various chemical composition and *in vitro* gas production parameters of the tannin rich tree leaves is given in Table 6. Significantly negative correlation was obtained between TT or CT and OMD, and DMD (r = - 0.454 to - 0.729; *P*<0.01). Similarly, CT was negatively correlated with gas production $\geq$ 12 h and the potential extent of gas production (*P*<0.05).

Table 6. The correlation coefficient between some chemical composition and *in vitro* fermentation parameters for tannin rich tree leaves

	CT	aNDF-NDF	ADF-ADF	Lign	OMD	DMD	DCHO	TT	G_{12}	G_{24}	G_{32}	G_{48}	G_{72}	G_{96}	c	b
CT	1															
aNDF-NDF	0.187	1														
ADF-ADF	0.166	0.835**	1													
Lign	-0.281*	0.279*	0.336**	1												
OMD	-0.722**	-0.017	0.03	0.094	1											
DMD	-0.669**	0.08	0.144	0.085	0.931**	1										
DCHO	-0.454**	-0.083	-0.137	-0.552**	-0.294*	-0.264*	1									
TT	0.849**	0.113	0.024	-0.364**	-0.694**	-0.629**	-0.729**	1								
G_{12}	-0.264*	-0.104	-0.252	0.065	0.383**	0.283*	0.004	-0.246*	1							
G_{24}	-0.365**	-0.213	-0.317*	0.009	0.533**	0.384**	0.036	-0.311*	0.942**	1						
G_{32}	-0.362**	-0.251	-0.329*	-0.058	0.545**	0.384**	0.074	-0.295*	0.884**	0.981**	1					
G_{48}	-0.369**	-0.243	-0.266*	-0.104	0.541**	0.431**	0.054	-0.329*	0.79**	0.924**	0.972**	1				
G_{72}	-0.331**	-0.258*	-0.257*	-0.181	0.534**	0.433**	0.08	-0.317**	0.711**	0.866**	0.929**	0.984**	1			
G_{96}	-0.239*	-0.219	-0.201	-0.265*	0.498**	0.427**	0.239	-0.202*	0.723**	0.834**	0.882**	0.92**	0.929**	1		
c	0.514**	-0.21	-0.425	-0.415*	-0.455**	-0.755**	0.345*	0.511**	-0.098	-0.266	-0.218	-0.345*	-0.426*	-0.319	1	
b	-0.594*	0.085	0.095	-0.112	0.065	0.154	0.45**	-0.13*	0.039	0.03	0.011	-0.014	-0.015	0.105	0.099	1

*CT, condensed tannin; aNDF-NDF, neutral detergent fiber assayed with heat stable alpha amylase and expressed inclusive of residual ash; ADF-ADF, acid detergent fiber expressed inclusive of residual ash; Lign, acid detergent lignin; OMD, organic matter digestibility; DMD, dry matter digestibility; DCHO, digestible carbohydrate; TT, total tannin; G_{12}, gas production at 12 h incubation; G_{24}, gas production at 24 hour incubation; G_{32}, gas production at 32 hour incubation; G_{48}, gas production at 48 hour incubation, G_{72}, gas production at 72 incubation; G_{96}, gas production at 96 hour incubation; 'c', rate constant of gas production during incubation (%/h); 'b', the potential extent of gas production; **P<0.05; **P<0.01.*

Discussion

Plant leaf effects on relative nutritive value

The CP content calculated for leaves of the MPTS in this study is much higher than the minimum CP level (70 g CP/kg DM) required for optimum functioning of rumen and for adequate intake of forages (Norton, 2000; McDonald et al., 2002). The CP content of all plant leaves was also far above the 60 g CP/kg DM level recommended as minimum daily requirement of ruminant animals from tropical feeds (NRC, 1984).

The threshold level of NDF in tropical plants beyond which DM intake and digestibility of ruminants affected is 600 g NDF/kg (Meissner et al.,1991). This suggests that leaves of all MPTS in the present study score well in terms of NDF values (below 600 g NDF/kg DM) and this indicates that the studied MPTS have high feeding value in terms of NDF content. Although the detergent fibre extraction techniques are used regularly when analyzing foliage from trees and other browse plants, Balogun et al (1998) and Makkar et al. (1995) pointed out that these techniques are not suitable for tanniniferous forages. Therefore, these values might need to be considered with caution.

As nitrogen utilization by rumen microbiota is related to the amount of fermentable energy, the adequate NSC contents (not less than 200 g/kg DM at matured leaves) (Marais et al., 2000; Barrau et al., 2005; Tylutki et al., 2008) especially in the leaves of MPTS species could enable efficient microbial protein synthesis by promoting better utilization of rumen ammonia nitrogen (NH_3-N) released from feeds with high content of rumen degradable CP (Cabrita et al., 2006).

The CT contents were highly variable with significant ($P<0.001$) differences among the species and more than 50 g/kg DM in all species. The effect of tannins can be either adverse or beneficial for animals, depending on the concentration and chemical structure (Makkar 2003; Min et al 2003). It has been also reported that CT values above 50 g/kg DM can become a serious anti-nutritional factor in plant materials fed to ruminants and are even toxic (Leng, 1997). Barry and Manley (1984) explained that a forage containing more than 50 g CT/kg DM is considered tannin-rich forage. Higher tannin levels (CT>50 g/kg DM) become highly detrimental (Barry and Manley, 1984) as they reduce digestibility of fiber in the rumen by inhibiting the activity of bacteria (Chesson et al., 1982) and anaerobic fungi (Akin and Rigsby, 1985), high levels also lead to reduced intake (Leng, 1997). Brooker et al (1999) also reported that livestock animals consuming tannin-rich diets over 50 g CT/kg DM usually develop a negative nitrogen balance and lose weight and body condition unless supplemented with non-protein nitrogen, carbohydrate and minerals. Silanikove et al. (2001) and Makkar (2003) suggested the dietary inclusion of tannin deactivating agents in order to improve utilization of tannin rich diets.

A feed containing over 700 g digestible DM/kg is defined as high quality (Meissner et al., 1991), however, the digestible DM of leaves of all MPTS were below the recommended value. The lower DM digestibility (<700 g/kg DM) of the MPTS could be due to the presence of CT which inhibited normal action of the rumen microbial enzymes (Moore and Jung, 2001; SCA, 1990; Alimuddin, 2007). In general, variations in chemical composition, tannin concentration and digestibility among different leaves of the MPTS may be partly due to genotypic factors that control accumulation of plant nutrients.

In vitro gas production potentials of tannin rich tree leaves

The gas production value more than 100 ml was measured at 12 h incubation for only two species whereas 60% and 90% of the species had a three digit gas yield (>100ml) after 24 and 32 h of incubation, respectively. This can indicate that the majorities of the MPTS attain optimum gas production after 24 and 32 hours of incubation. The organic matter and metabolisable energy value estimation based on 24h gas production (Menke and steingass, 1988) could not apply for most of the MPTS since the best possible gas production cannot attain at 24 hour. Tanniferous plants have slower fermentation capacity (Getachew et al, 2001, 2002). Hence, many laboratories adopted a 16-48 h gas production protocol for valid measurements of digestibility and metabolisable energy from rapidly fermentable substrates as well as slowly fermentable substrates. According to Menke and Steingass (1988), Theodorou et al (1994) and Groot et al (1996) the gas production from both rapidly and slowly fermentable substrates can be precisely measured considering those incubation hours mentioned above.

Gas production was found to be not consistent with the chemical composition of the leaves of the MPTS species. This could be due to the binding effect of CT on proteins that latter reduces fermentation ability of browse sources. Condensed tannins could play vital role in reducing initial gas production in tannin rich species by binding available proteins and other essential nutrients for fermenting microbes. A report by Kamalak et al (2004b) revealed that *in vitro* gas production and polyphenolic tannin concentrations were negatively correlated showing inverse linear relationships with either the tannin concentrations or their biological activity.

Effects of tannin binding agents on in vitro fermentation of leaves of tannin rich trees

The significant improvement in *in vitro* gas production(Table 3) of all MPTS leaves following incubation with TBAs (PEG_{4000}, PEG_{6000}, PVP and PVPP) might indicate that tannin deactivating ability of TBAs by forming inert tannin deactivating agent: tannin complexes, which latter result in increased in gas production. Moreover, an increment in gas production after inclusion of TBAs clearly shows tannins decrease accessibility of proteins to rumen microbes by binding fermentable proteins. Reports by Getachew et al (2001), Seresinhe and Iben (2003) and Singh et al. (2005) revealed that incubating tannin-containing feed resources with PEG_{4000} improved *in vitro* gas production. However, no report has been compared for the efficacy of PEG_{6000}, PVP and PVPP on *in vitro* gas production of leaves of tannin rich trees in another place. Tannins may form a less digestible complex with dietary proteins and may bind and inhibit the endogenous protein, such as digestive enzymes which, can reduce *in vitro* fermentation ability. Moreover, increases in gas production due to incubation with tannin deactivating agents could be due to an increase in SCFA production and/or changes in molar proportions of the short chain fatty acids. Getachew et al (2002) and Makkar (2003) reported the positive association between SCFA and *in vitro* gas production. Thus, in the present study it has clearly marked that removal of adverse effects of tannins by TBAs increased the fermentability and energy content of the MPTS. The significant increment in *in vitro* gas production in the presence of TBAs might be associated with increasing availability of nutrients to rumen micro-organisms, especially the available nitrogen. After incubation of substrates with TBAs, in the present study we demonstrated a larger increase in gas production which is irrespective of CT concentration in all MPTS. In the current study, both types of PEGs (PEG_{4000} & PEG_{6000}) produced significantly larger amounts of gas when compared with other tannin binding agents (P<0.001). This might be associated with their lower molecular weight as compared to PVP and PVPP. Makkar (2003) indicated the direct relationship between molecular weight, solubility and permeability of TBAs and *in vitro* fermentation. Although PEG_{4000} has been the most widely used TBA, in the present study it was observed that PEG_{6000} was more effective in overcoming the effects of tannins at neutral pH *in vitro*. It is likely to have specific interactions with tannins compared to other TBAs in closely neutral pH. TBA: tannin interaction is pH dependent (Silanikove et al., 2001, Makkar, 2003).

The present study revealed that the gas production was highly associated with the dosage of tannin binding agents, meaning that the higher the dosage of TBAs, the higher the volume of gas production was measured. Nevertheless the antinutritional properties of tannins depend on their chemical structure and dosage, which can imply that sufficient dosages of TBAs are required to cleave the bond tannin: substrate, hence allowing the animals to make use of essential nutrients for their physiological process.

Effect of plant species variation on in vitro fermentation kinetics

The *in vitro* fermentation kinetics of tannin rich tree leaves is presented in Table 4. The significant variation *in vitro* fermentation kinetics between plant species might be due to differences in fermentation patterns of plant species. Kafilzadeh and Maleki (2012) observed differences in asymptotic gas production of plant leaves due to differences in plant species and/or variety. The higher values obtained for the potential extent of gas production b in the *G. ferruginea, P. africana* and *S. guineense* indicating that these plants have better nutrient availability for rumen microorganisms. The higher b value recorded for *E. capensis* and *M. ferruginea* is associated with their lowesr CT, yet, the potential gas production in *F. sycomorus* is irrespective of its CT concentration. This might be associated differences in type of tannin, which is not measured in this research, as well as differences in chemical composition of the plant species. These differences are essential for an examination of end product kinetics at different time points in relation to maximum substrate disappearance. Hence, the intake of a feed is mostly explained by the rate of gas production 'c' which affects the passage rate of feed through the rumen, whereas the potential extent of gas production b is associated with degradability of feed (Khazaal et al., 1995).

Effect of tannin binding agents on in vitro fermentation kinetics of tannin rich tree leaves

Incubation of tannin rich tree leaves with PEG_{4000}, PEG_{6000}, PVP and PVPP had increased potential extent of gas production b over control treatments whereas incubation of the leaves with tannin binding agents induced a decrease in rate of gas production in all plant species (Table 5). The significant improvement in the fermentation kinetics of tannin rich tree leaves after incubation with TBAs could be associated with the non-restricted colonization of the leaf substrates by fermenting bacteria and/or other beneficial microbes. The complex forming between tannins and PEG_{4000}, PEG_{6000}, PVP or PVPP could generate sterric obstructions which do not limit the fixation of adherent bacteria to the feeds. Similar to potential gas production values (Table 3), the larger potential extent of gas production b after PEG_{6000} and PEG_{4000} and their larger doses as compared to other TBAs could indicate the suitability of these TBAs for *in vivo* studies; however, the efficacy of the TBAs is not clearly associated with the respective

CT concentration in the tree leaves. Further study on tannin type and structure might give a clear effect of TBAs on tannins. The PEG$_{4000}$ supplementation (Canbolat et al., 2005) increased potential extent of gas production b, but no effect on the gas production rate c.

Correlation coefficient between some nutritive value and in *vitro gas* production parameters

The correlation coefficient between DMD, OMD and *in vitro* gas production was found to be positive and significant for the various incubation hours ($P<0.01$) (Table 6). A possible reason to this fact is that gas production is an end product of digestible and potentially fermentable nutrients. The negative correlation coefficient between digestible nutrients and tannin concentrations showing an inverse linear relationship with either the tannin concentrations or their biological activity assessed using tannin binding agents. The findings in this experiment supported the fact that the anti-nutritive factors like tannins may also contribute to reduction of microbial activity. In contrast, Larbi et al (1998) and Abdulrazak et al (2000) reported a weak relationship between CT and gas production parameters of the tree leaves during wet and dry season in West Africa. A possible reason could be due to differences in the nature of tannins between browse species. The negative correlation coefficient between potential extent of gas production b and CT might be associated with the reduction of microbial activity from increasingly adverse environmental conditions as incubation time progress where as the positive correlation between b and DCHO could imply the roles of digestible carbohydrates as energy source for fermenting rumen bacteria.

Conclusion

The evaluated tannin rich tree leaves showed high variations in chemical composition, digestible nutrients, ME, and extents and kinetics of *in vitro* gas production characteristics. Although the evaluated fodder tree leaves seem appropriate as such for ruminant feeding in tropical conditions, the addition of TBAs such as PEG_{4000}, PEG_{6000}, PVP and PVPP makes a considerable improvements in nutritive value. Among the TBAs included in this study both types of PEG rendered the highest *in vitro* gas production followed by PVP and PVPP. Moreover, increasing the inclusion level of the tannin binding agents has revealed a significant increase in gas production, and kinetics of fermentation. Yet, dosage and types of TBAs have to be considered, and the TBA effect size was not clearly related to the CT concentration in the plant leaves. Since this study only covered ten new tree species rich in tannins, it is recommended that further studies should be carried out to evaluate more locally available browse species so that the feed value of the new potential plant species can be known. Animal feeding experiments are also necessary to ascertain the nutritive value of such the promising plant species with or without TBAs.

Acknowledgement

The authors gratefully acknowledge the VLIR- UOS institutional university cooperation program (IUC-JU) for funding the research budget.

References

Abdulrazak S.A., Orden E.A., Ichinohe T., & Fujihara, T. (2000). Chemical composition, phenolic concentration and in vitro gas production characteristics of selected Acacia fruits and leaves. *Asian-Aust. J. Anim Sci.* 13, 935-940.

Aberra, M. (2012) Assessing the feeding values of leaves, seeds and seeds-removed pods of *Moringa stenopetala* using *in vitro* gas production technique. *African J Biotechnol* 11(51),11342-11349.

Akin D.E., & Rigsby LL (1985). Influence of phenolic acids on rumen fungi. Agronomy Journal, 77:180-182.

Alimuddin N. (2007). Animal nutrition training manual. Pakistan National Animal Husbandry. AKF Kabul – Afghanistan, pp 77.

AOAC (Association of Official Analytical Chemists) (2005). Official Method of Analysis. 15th ed. Washington, DC., USA. pp. 66–88.

Arigbede O.M., Anele Südekum K.H., Hummel J Oni AO Olanite J.A. & Isah O.A. (2011). Effects of species and season on chemical composition and ruminal crude protein and organic matter degradability of some multi-purpose tree species by West African dwarf rams. *J Animal Physiol Animal Nutrit.* 96(2), 250-9.

Babayemi O.J. (2007). *In vitro* fermentation characteristics and acceptability by Wes African dwarf goats of some dry season forages. *African J. Biotech.* 6(10), 1260-1265.

Balogun R.O., Jones R.J.& Holmes J.H.G. (1998). Digestibility of some tropical browse species varying in tannin content. Anim feed Sci Technol 76:77-88.

Barrau E. Fabre N., Fouraste I. & Hoste H (2005). Effect of bioactive compounds from sainfoin (Onobrychis viciifolia Scop.) on the in vitro larval migration of *Haemonchus contortus*: role of tannins and flavonol glycosides. *Parasitology* 131,531–538.

Barry, T.N. & Manley, TR (1984). The role of condensed tannins in the nutritional value of *Lotus pedunculatus* for sheep. Quantitative digestion of carbohydrates and proteins. British J Nutrit. 51,493-504.

Besharati, M. & Taghizadeh, A. (2009). Evaluation of dried grape by-product as a tanniferous tropical feedstuff. *Animal feed Sci Technol.* 152(3-4):198–203.

Blümmel, M. & Ørskov, E.R. (1993). Comparison of gas production and nylon bag degradability of roughages in predicting feed intake in cattle. *Animal Feed Sci Technol.* 40, 109-119.

Blummel, M., Aiple, K.P., Steingass, H. & Becker K (1999). A note on the stoichiometrical relationship of short chain fatty acid production and gas formation *in vitro* in feedstuffs of widely different quality. *J Animal Physiol Animal Nutrit.* 81, 157-167

Brooker, J.D, O'Donovan, L., Skene I. & Sellick G (1999). Mechanism of tannin resistance and detoxification in the rumen. Pp. 117-122. In: tannin in livestock and human nutrition (ed. Brooker), proceeding of an international workshop held in Adelaide, Australia, 31 May-2 June 1999. Australian Center for International Research (ACIAR). Canberra, Australia.

Cabrita A.R., Dewhurst, J.R.J, Abreu J.M. F., Fonseca, A. J. M. (2006). Evaluation of the effects of synchronizing the availability of N and energy on rumen function and production responses of dairy cows- review. *Animal Res.* 55,1-24.

Canbolat, O., Kamalak, A., Ozkose, E., Ozkan C.O, Sahin, M. & Karabay, P. (2005). Effect of polyethylene glycol on *in vitro* gas production, metobolizable energy and organic matter digestibility of Quercus cerris leaves. *Livest Res Rural Develop.* 17(4).

Cerrillo, M.A. & Juarez, R.A.S. (2004) In vitro gas production parameters in cacti and tree species commonly consumed by grazing goats in a semi arid region of North Mexico. *Lives. Res. Rural Dev.* 16(4), 9.

Chenost, M., Deverre, F., Aufrere, J. Demarquilly, C (1997). The use of gas-test technique for predicting the feeding value forage plants. In: *In vitro* techniques for measuring nutrient supply to ruminants. Proceedings of Occasional Meeting of the British Society of Animal Science, 8-10 July 1997, University of Reading, UK.

Chesson, A., Stewart, C.S & Wallace R.J. (1982). Influence of plant phenolic acids on growth and cellulolytic activity of rumen bacteria. Appl. Environ. *Microbiol.* 44,597–603.

Deckers S., Van Ranst E. Poesen, J. (2008). Report on reconnaissance survey. Soil Fertility Project, VLIR JU-IUC.

El Hassan Kassi & Newbold Wallace (2000). Chemical composition and degradation characteristics of foliage of some African multipurpose trees. *Animal Feed Sci Technol.* 86, 27-37.

FAO/IAEA, (2000) Quantification of tannins in tree foliage. IAEA-TECDOC, IAEA, Vienna, Austria

Frutos, P., Hervás, G, Giráldez, F.J & Montecón A. R (2004). An in vitro study on the ability of polyetylene glycol to inhibit the effect of quebracho tannins and tannic acid on rumen fermentation in sheep, goats, cows and deer. *Australian J Agri Res* 55, 1125-1132.

Getachew, G. Makkar H.P.S., Becker K. (2001). Method of polyethylene glycol application to tannin-containing browses to improve microbial fermentation and efficiency of microbial protein synthesis from tannin-containing browses. *Animal feed sci Technol.* 92, 51-57.

Getachew, G., Makkar H.P.S.& Becker, K. (2002). Tropical browses: content of phenolic compounds, in vitro gas production and stoichiometric relationship between short chain fatty acids and in vitro gas production. *J Agri Sci.* 139,341.

Getachew G Robinson PH de Peters EJ and Taylor SJ (2004). Relationships between chemical composition, dry matter degradation and *In vitro* gas production of several ruminant feeds. *Animal Feed Sc Technol.* 111, 57-71.

GOR (Government of Oromia Region, Ethiopia) (2006). Socio-economic profile of the Jimma zone. Federal Ministry of Health of Ethiopia.

Groot JCJ Cone J Williams BA Debersaques F D and Lantinga EA (1996). Multiphasic analysis of gas production kinetics for in vitro fermentation of ruminant feeds. Animal feed Sci Technol 64,77-89.

Hagerman Ann E (2011). The Tannin Handbook. http://www.users.muohio.edu/hagermae

Hveplund T Moller PD Madsen J and Hesselholt M (1995). Flow of digesta through the gastrointestinal track in the bovine with special reference to nitrogen. In: year book, Royal vet Agr Univ, Copenhagen, pp.173-192.

Kafilzadeh F and Maleki E (2012). Chemical composition, in vitro digestibility and gas production of straws from different varieties and accessions of chickpea. Department of Animal Science, Faculty of Agriculture, Razi University, Kermanshah, Iran. *J. Animal Physiol. Animal Nutrit.*. 96 (3), 450-8.

Kaitho RJ Tegegne A Umunna N N Nsahlai I V Tamminga S van Bruchem J and Arts JM (1998). Effect of *Leucaena* and *Sesbania* supplementation on body growth and scrotal circumference of Ethiopian highland sheep and goats fed tef straw basal diet. *Lives. Prod. Sci.* 54,173-181.

Kamalak A Canbolat O Ozay O and Aktas S (2004b). Nutritive value of oak (*Quercus spp.*) leaves. *Small Ruminant Res.* 53(1-2), 161-165.

Khazaal K A Dentino M T Ribeiro J M and Ørskov E R (1995). Prediction of apparent digestibility and voluntary intake of hays fed to sheep: comparison between using fibre components, in vitro digestibility or characteristics of gas production or nylon bag degradation. *Anim.Sci.* 61, 527-538.

Larbi A Smith JW Kurdi IO Adekunle IO Raji A M Ladipo DO (1998). Chemical composition, rumen degradation, and gas production characteristics of some multipurpose fodder trees and shrubs during wet and dry seasons in the humid tropics. *Animal Feed Sci Technol.* 72, 81-96.

Leng RA (1997). Tree Foliage in Ruminant Nutrition. FAO Animal Production and Health Paper. No. 139. FAO (Food and Agriculture Organization), Rome, Italy.

Makkar H P S Borowy N Becker K Degen R (1995). Some problems in fiber determination of a tannin-rich forage (Acacia saligna leaves) and their implication in *in vivo* studies. *Animal Feed Sci Technol.* 55,67-76.

Makkar HPS (2003). Effects and fate of tannins in ruminant animals, adaptation to tannins, and strategies to overcome detrimental effects of feeding tannin-rich feeds. *Small Rum Res.* 49, 241–256.

Makkar H P S and Becker K (1998). Do tannins in leaves of trees and shrubs from Africa and Himalayan regions differ in level and activity? *Agroforestry Syst.* 40, 59–68.

Makkar H P S Blümmel M Borowy N K and Becker K (1993). Gravimetric determination of tannins and their correlations with chemical and protein precipitation methods. *J. Sci. Food Agric.* 61, 161–165.

Marais JPJ Mueller-Harvey I, Brandt EV and Ferreira D (2000). Polyphenols, condensed tannins, and other natural products in *Onobrychis viciifolia* (sainfoin) J Agri Food Chem. 48, 3440–3447.

McDonald P Edwards AR Greenhalp JFD, Morgan CA (2002). Animal nutrition (6th ed.). Prentice Hall. London.

Meissner HH and Paulsmeier DV (1995). Plant composition constituents affecting between-plant and animal species prediction of forage intake. J *Animal Sci.* 73, 2447-2457.

Meissner HH, Viljoen MO, Van Niekerk WA (1991).Intake and digestibility by sheep of Antherphora, Panicum, Rhodes and Smooth finger grass. Proceedings of the 4th International Rangeland Congress, September 1991, Montpellier, France, pp: 648-649.

Menke KH and Steingass H (1988). Estimation of the energetic feed value obtained from chemical analysis and in vitro gas production using rumen fluid. *Animal Res Deve* 28, 7–55.

Min B Barry T Attwood T and McNabb W (2003). The effect of condensed tannins on the nutrition and health of ruminants fed fresh temperate forages: a review. *Animal Feed Sci Technol.* 106, 3-19.

Moore KJ and Jung HJG (2001). Lignin and fiber digestion. J Range Manag 54, 420–430.

Njidda AA and Nasiru A (2010). *In vitro* gas production and dry mater digestibility of tannin-containing forges of semi-arid region of north-eastern Nigeria. *Pakistan J Nutrit* 9, 60-66.

Norton B W 2000 The significance of tannins in tropical animal production. In Tannins in livestock and human nutrition. Proceedings of an International Workshop (ed. JD Brooker), pp. 14–23. Australian Centre for International Agricultural Research, Canberra, Australia.

NRC (National Research Council) (1984). National Research Council: Nutrient requirements of domestic animals: goats: National Academy Press. Washington DC.

Okoli IC, Anunobi Maureen O, Obua BE, Enemuo V (2003). Studies on selected browses of southeastern Nigeria with particular reference to their proximate and some endogenous anti-nutritional constituents. *Lives. Res. Rural Dev.*15(9).

Ørskov ER and McDonald Y (1979). The estimation of protein degradability in the rumen from incubation measurements weighed according to rate of passage. *J. Agri. Sci.* 92, 499–503.

Ozturk D Kizilsimsek M Kamalak A Canbolat O and Ozkan CO (2006). Effects of ensiling alfalfa with whole-crop maize on the chemical composition and nutritive value of silage mixtures. *Asian-Australian J Animal Sci* 19, 526-532.

Porter LJ Hrstich LN and Chan BG (1986). The conversion of pro-cyanidins and prodelphinidins to cyaniding and delphinidin. *Phytochemistry*, 25, 223-230.

Ranjhan SK (1997). Animal Nutrition in the Tropics. 4th edition, Modern Printers, Delhi, pp.59-99.

Reed, J.D., McDowell, R.E., Van Soest, P.J. & Horvath P.J. (1982). Condensed tannins, a factor limiting the use of cassava forage. *J Sci Food Agri* 33, 213-220

Robertson, J.B. & Van Soest P. J. (1981). The detergent system of analysis. In: James, W.P.T., Theander, O. (Eds.), The Analysis of Dietary Fibre in Food. Marcel Dekker, N.Y.,Base, L., pp. 123, 158.

Robinson, P.H., Wiseman, J., Udén, P. & Mateos, G G (2006). Some experimental design and statistical criteria for analysis of studies in manuscripts submitted for consideration for publication. *Anim. Feed Sci. Technol.* 129, 1–11.

Salem, A.Z.M., Salem, M.Z.M., El-Adawy, M.M.& Robinson, P.H. (2006). Nutritive evaluations of some browse tree foliagesduring the dry season: Secondary compounds, feed intake and in vivo digestibility in sheep and goats. *Animal Feed Sci. Technol.* 127, 251–267.

Sallam, S.M.A, Nasser, M.E.A., El-Waziry, A.M, Bueno, I.C. S.& Abdalla, A.L. (2007). Use of an *in vitro* Rumen Gas Production Technique to Evaluate Some Ruminant Feedstuffs. *J Applied Sci Res.* 3(1), 34-41.

SAS (2010). Statistical Analysis System, Release 9.3 SAS/STAT Program. SAS Institute Inc., Carry, NC, USA.

SCA (1990). Feeding Standards for Australian Livestock. CSIRO Publications, East Melbourne, Australia.

Seresine T and Iben C (2003). *In vitro* quality assessment of two tropical shrub legumes in relation to their extractable tannins content. *J Animal Physiol Animal Nutrit.* 87,109-115.

Silanikove, N. Perevoltsky, A. & Provenza FD (2001). Use of tannin binding chemical to assay for tannin and their negative postingestive effects in ruminants. *Feed sci. technol.* 91, 69-81.

Singh, B., Sahoo, A., Sharma, R. & Bhat, T.K. (2005)). Effect of polyethylene glycol on gas production parameters and nitrogen disappearance of some tree forages. *Animal Feed Sci. Technol.* 123-124, 351-364.

Theodorou, M.K, Williams BA, Dahona MS, McAllan A.B. & France, J. (1994). A simple gas production method using a pressure transducer to determine the fermentation kinetics of ruminant feeds. *Animal Feed Sci Technol.* 48, 185-197.

Tilley, J.M.A. & Terry R. A. (1963). A two-stage technique for the in vitro digestion of forage crops, J British Grass land Soci.18, 104-111.

Tylutki, T.P Fox, D.G, Durbal, M., Tedeschi L.O., Russel J.B., Van Amburgh, M.E., Overton, T.R., Chase L.E. & Pell A.N. (2008). Cornell Net Carbohydrate and Protein System: A model for precision feeding of dairy cattle. *Animal Feed Sci Technol.* 143,174-202.

Van Soest, P.J. & Robertson, J.B. (1985). Analysis of forages and fibrous foods. A Laboratory manual for animal science, 613, Cornell University, USA

Van Soest, P.J., Roberston J.B. & Lewis, B.A. (1991). Methods for dietary fibre and nonstarch polysaccharides in relation to animal nutrition. *J Dairy Sci.* 74,3583-3597.

Van Soest, P.J. (1994). Nutritional ecology of ruminants, 2nd edition, Cornell University Press

Yisehak, K., Belay, D. & Janssens, G.P.J. (2010). Indigenous fodder trees and shrubs: nutritional and ecological aspects. The 15th congress of ESVCN, 15th of September, 2010. Zurich, Switzerland.

YOUR KNOWLEDGE HAS VALUE

- We will publish your bachelor's and master's thesis, essays and papers

- Your own eBook and book - sold worldwide in all relevant shops

- Earn money with each sale

Upload your text at www.GRIN.com and publish for free